MORE SILENCE
LESS NOISE

WHEN ENVIRONMENTAL NOISE

DISTURBS YOUR LIVING

By

I0478405

PETER KRUSE, MD, PHD

Edition 1.0

Cover layout and photo: Irene Livia

ISBN-13: 978-1546942498

ISBN-10: 1546942491

ABBREVIATIONS

°C	Degrees celcius
DALY	Disability-adjusted life-years
dB	Decibel
DJ	Disc jockey
EU	European Union
Hz	Hertz
M	Meter
NRR	Noise reduction rating
Rpm	Revolutions per minute
TV	Television
US	United States
WHO	World Health Organization

INTRODUCTION

If you prefer silence over noise - and value silence
as an important part of your life – then perhaps
this book is written for you.

It took me some years of living, perhaps 25 years,
before I understood that I preferred silence over
noise. I actively sought to avoid noise in my daily
life and even later in life, at the age of 40, I began
to plan my activities in ways that favored silence . I
found noise to have a negative impact on my life; it
distracted and drained me.

Since my birth in the beginning of the sixties, I
have experienced a world where noise has
become even more widespread due to a lot of
technological evolutions. Take as an example the
"wearable" music that began really with the
Walkman: Now you could listen to music
anywhere and as you could hear in a song from
the 70': "Now I have bought myself a Walkman
and now the forest sounds like New York – man!"
Today the same wear ability can be found through
music in smartphones, smart watches etc. But
music is far from the only environmental noise
source: Traffic, construction, machines of every

kind, public spaces, houses, work places, ways of travel, pets and many more. A lot of modern technology helps us by generating noise such as alarms, beeps, signals etc. All this noise is part of "Environmental Noise" that has an impact on our lives. It may even have a very negative impact on all phases of our lives from early learning, studying in school or university, work/life balance, everyday interactions, friends, family, sleep, love life, health and much more.

Perhaps the technological revolution that led to the generation of many more noise sources has run too fast and out of control so we are not anymore able to manage and minimize noise in our environment?

We could learn something from the evolution of smoking, another environmental polluter: Now we can read anywhere how bad smoking is. Back in the sixties smoking was generally accepted in places such as restaurants, trains, airplanes etc. It was unthinkable to have restrains on where you could smoke. Even in airplanes where many people were in the same air cabin for hours smoking was allowed. Today science has clearly shown that smoking is bad for your health and on all tobacco products you will find the label "Smoking kills" and a total ban on smoking in public places. Could there be a similar need to revisit the negative impact of noise on our lives and to act upon that as with smoking? World Health Organization (WHO) - the lead expertise in global health issues including environmental noise - estimates that more than one million healthy life years are lost every year from traffic-related noise in western Europe. So perhaps it is not only the author that is disturbed from the noise generated in the environment around us.

This book is intended to achieve a knowledge on the impact of environmental noise on our everyday life including on our health. The health aspects are indented to be evaluated from a scientific point of view. I view noise as a general problem in any society and it is without any borders. As such, this book intends to raise a broad public debate on how we as individuals can develop a noise free environment. Such a debate must involve politicians and needs to change laws and norms to achieve its broad purpose.

We need to act <u>together and quickly</u> towards a vision of:

A society where silence is prioritized over noise.

I hope you share that vision.

Peter Kruse, MD, PhD, June 2017

Table of Contents

LEGAL NOTES

MORE SILENCE LESS NOISE

WHEN ENVIRONMENTAL NOISE DISTURBS YOUR LIVING

By PETER KRUSE, MD, PHD
Copyright © @ 2017

CHAPTER 1. SILENCE VERSUS NOISE

Silence is defined as "*absence of any sound or noise; stillness*"[1]. Sound is vibrations moving as waves of pressure through air or water. The human ear will be able to hear sound waves typically with the frequencies between 20 Hz and 20.000 Hz. Animals may hear different frequencies than humans.

It is of course difficult to find places with absolute silence (no sound at all) outside perhaps specially built soundproof room where no noise enters or escapes the room. In the nature, a lot of sound is generated from the weather such as wind, rain, and thunder and from the animals such as birds and insects etc. This sound from the nature (ambient sounds) is often very calming to listen to. Wind blowing in the threes or the sound of the ocean waves crushing onto the shore – provides your mind with energy. But the nature can be really silent. It is a pleasure going to the mountains where on a day with no wind or rain you can enjoy the silence of the splendid nature. On days without any wind, the first thing you notice if you take the ski-lift to the mountain top is the sudden silence that surrounds you. Total silence in the nature is rare: But it is possible to experience the nature's total silence – and the author has tried this: On a clear day without any clouds you could be standing in the middle of the Inland ice in Greenland and you will see only two colors: Blue sky above and white snow below. You will be overwhelmed by the incredible silence. You hear nothing, absolutely nothing. If you move the booths on the snow the quirky sound seems to be absorbed by this enormous silence. It is very energizing experience and you will remember this silence for the rest of your life.

Erling Kagge who is a polar explorer and author of "Silence: In the age of noise" described his meeting with the silence in Antarctica in a beautiful way[2]. Here he explains that in the beginning of his journey the color of the snow was white but with the silence his senses sharpened and he was able to detect the numerous shades of the colors of the snow that surrounded him. Perhaps what he is intending to tell us is that the noise around us "numbs" our senses so we are unable to detect and enjoy the nuances.

But silence can also be enjoyed in a man-made environment. One such place is a Japanese rock garden (or a *karesansui*); a garden of rocks, gravel, sand and perhaps small bonsai trees. Silence here is a key element in visiting a karesansui. With silence you are able to see details of the rock garden that noise would have distracted you from. A place for meditation.

Nearly all religious sacred places such as mosques, churches, shrines, synagogues call for silence when you enter such a place. Silence helps you focus – helps you see better and listen better and perhaps even think better about whatever religious thoughts you may have. It feels natural for us visitors of such sacred places to respect the silence and perhaps also to enjoy it.

When the auditory input disappears you sharpen your other senses. You now have full focus on the colors of the leaves or the shades in the sand dunes. You observe better in silence. If you visit the beautiful Guggenheim museum in Venice you may observe all the other visitors from a quiet spot in the garden – perhaps you will be able to enjoy that a bit more than the actual art in the museum as here it is very noisy. Silence sharpens your other senses.

From a state of silence we humans will perceive added sound individually to a point where the sound will become annoying. This is where the definition of "noise" begins. It is important to understand that the perception of when sound becomes annoying (noise) is very subjective. You may work with the TV on without feeling this as a disturbing noise while another person will have to turn off the same TV with the same volume as it disturbs her/him.

When sound actually becomes annoying is very subjective

With noise this book will only deal with environmental noise defined by the World Health Organization (WHO) as "noise emitted from all sources except noise at the industrial workplace"[3]. Environmental noise include traffic, industry, construction and the neighborhood as the main sources[3]. Whether specific sounds are perceived as noise by an individual will depend on many different factors such as[3]:

- Sound pressure (level of air vibrations that make up sound)

  The sound pressure level is the logarithmic measure of the effective pressure of a sound relative to a reference value and is measured in decibel or dB. Here are some examples of sound pressure levels in dB:

Noise source	Sound pressure level (dB)
Threshold for hearing of the normal human ear (at 1000 Hz)	0
Very calm room	20-30
Normal conversation (1 m)	50
TV listening at home (1 m)	60
Traffic on a busy road (10 m)	85
Road construction with Jack hammer (1 m)	100
Jet engine (100 m)	120-140

- Frequencies in Hz

  Frequencies outside what the human ear can perceive can still be sensed by the human body either as pulses (low frequencies from a diesel engine) or as a pitch (high frequencies from a dog whistle).

- Continuous noise

  Continuous sound pressure like the one from a heavily trafficked road.

- Peaked noise

  Gun shots produce high peaks of sound pressure with perhaps complete silence between the peaks.

In the "Guidelines for community noise"[3] written by the WHO the complexity of measuring noise annoyance to people is well described. Factors such as noise variation with time, noise frequency and loudness, other competing ambient noise sources, and types of noise are all important in determining whether a noise is disturbing to the individual or not.

A lot of effort by politicians is put into funding research on how to measure noise and into mapping where noise occurs. That is necessary and good. However, in my opinion, and this is one of the main messages of this book, it is critical to understand that one of the most important factors in determining whether a sound becomes annoying to a person is her/his own subjective perception of the noise:

- **We all respond differently to noise**

This fact makes it difficult for the scientists to come up with a measure to quantify noise. Currently different weighting measures are used to measure noise from various sources (e.g., LAeqT, LAmax, or SEL)[3]. But none of these weights can include the human variability of how we differently respond to noise. In other words we have all different "sensitivity" to noise: If you measured the noise generated from our TV this would give some number of noise impact; your daughter would easily fall asleep in spite the noise and sleep well for hours while the TV in on – you couldn't do that; You would most likely need complete silence for falling asleep. Young people are nowadays used to studying with earphones listening to music while you would need complete silence while studying/reading/working.

The noise "sensitivity" is not only different between individuals, also within the same individual you may experience different sensitivity; age, the time of day (more sensitive at night), your activity when exposed to noise (how much do you need to concentrate), your willingness to be exposed to the noise (if you go to a concert your are prepared and want to hear the music), etc.

It is good that WHO[4], the EU Commission[5] and the U.S. Environmental Protection Agency[6] are trying to quantify and map the noise in order to generate guidelines for acceptable noise levels. However, our societies need to also include the individual different noise sensitivities into the equation:

- **People with low noise tolerance also have a right to live in an environment without for them disturbing noise**

Mapping of noise levels is done many places around the world. These maps can show you which streets have the highest noise levels. This quantitative noise mapping is excellent and should one day become part of Google maps so you can see where noise reduction activities are needed the most. However, people's perception of noise (qualitative noise mapping) is currently not being mapped i.e., where is noise an annoyance to people. That is something the future hopefully can bring.

CHAPTER 2. SNAPSHOTS FROM OUR LIFE WITH NOISE

When sound disturbs you it becomes noise.

So let us begin this book with an important fact:

Some of us are sensitive to noise.

We often notice that sound that disturbs one person does not disturb other persons around us when exposed to the exact same sound. We sense the noise that the others cannot hear/feel. Only when we insist: Are you sure you cannot hear that truck engine? Then the others we are with will admit that far far away they hear the sound but to them it is not annoying (as it is to us).

Silence gives us energy, peace, and clarity. We always seek silence over noise as we find silence much more rewarding to our body and mind. Through most of our life, silence has always been important to us and noise seems to have become a more profound "intruder" or "disturber" of our life over the years. We may have become more sensitive to noise over time or as an alternative, there is more noise around us – or maybe both.

Noise has had a negative impact on all aspects of our life. We do not sleep well with noise. Our reading and studying is distracting if we are disturbed by noise. Our work requires some degree of quietness to be able to concentrate and perform. Our relaxing time at home is much more relaxing with silence. Vacations are better if no noise is involved. And so on: All aspects of our life are "sensitive" to noise and to have a better life we have through the years learnt a way to cope with noise.

It is very hard to define when a noise becomes annoying to us. It depends a lot on the context in which the noise occurs and what we are doing when exposed to it. It also depends on the type and intensity of the noise. It is a bit like motion sickness: In motion sickness there is a mismatch between what the eye sees and what the vestibular system and nerve receptors inform your brain about where your body is in space. If you sit in a car or on a boat and you are looking at your smartphone your eyes do not see the movements that the car/boat is making – this gives you the nausea of motion sickness. In this case there is a mismatch between what your body feels and what your eyes see. It is something similar with the noise around us: If we are focused on the noise such as attending a music concert the sounds are not so annoying as if the same sounds occurred while working – here they would disturb our work as there is a "mismatch" between what we are doing and what we hear. So most like we have different noise thresholds depending on what situation we are in.

Some types of noise are rarely even detected by some people around us; only if we explain what they need to listen to can they (suddenly) hear this noise; one example of this is the very low noise generated from diesel engines such as in trains diesel locomotives or in large diesel trucks. This noise seems to penetrate walls very easily and is a kind of noise we may feel more than we actually hear it. Like a vibration that gives an uncomfortable sensation. Very loud and sudden noises are noise from barking dogs that provokes a more shock like reaction; a bit like what you see in small children exposed to a loud sudden noise: Sudden surprise and a fear like reaction. So what determines whether a noise disturbs us is in brief: Context, type, and intensity.

Let us look a couple of "snap-shots" taken from the author's real life experience with noise with different contexts, types, and intensities – perhaps you recognize some of these situations. These snap shots are meant as appetizers to illustrate what we may have begun to see as one of the major problems of our modern society that needs to be solved as soon as possible: Environmental noise (i.e., noise emitted from all sources excluding industrial workplaces).

Kids toy tractors: In an apartment the bedroom was facing the backyard. Perhaps six times a month you would have night shifts returning home at 9 o'clock in the morning after 24 hours of work. Obviously, you needed to rest but having been on the run all night made the falling asleep sensitive to (any) disturbance. You would eat some breakfast and with the curtains closed go to bed around 10 in the morning. This was the same time as the mothers in the apartment block would bring their small kids down to the backyard to play. Their favorite toy was a three wheeled plastic tractor with hard plastic wheels that when moved around in the backyard made an high pitched and intense noise on the yards tile flooring. This noise completely prevented you from falling asleep. Windows closed and ear-plugs – the special ones that have a noise reduction rating (NRR) of 30 dB were not enough to keep out the noise. The only way to solve this was for you to move to the living room to get some quietness and rest.

The festival: So the mayor wanted to attract people from outside the town to visit it by creating a festival. Food & drink stands and a lot of live music. From one big stage both live music and DJ's were ensuring music entertainment from late in the afternoon to one in the morning. Your home was located more than 500 meters from the festival stage. During the five days of festival the intense music was heard inside your home with closed windows to a degree that it was difficult to hear the own TV. Your family and several hundred others had to wait to go to sleep until one in the morning each of the five festival days. The interesting thing is that most of the festival guests came from outside of the town (which was the mayor's intention). But the goal was reached at the expense of the people living there due to the festival noise.

Diesel train engines: In an apartment close to the train station and tracks, the noise of the regularly passing trains was really not a problem. Modern train technology had really reduced the train noise. At night with the windows closed it was easy to sleep. However that changed: During winter time the cold was intense especially at night. With temperatures often below 0 °C and sometimes as low as -15 °C the railways decided to keep the old locomotives (diesel engines) running all night parked a few 100 meters from the actual train station. The sound of these diesel engines especially in idle state - where they do not move the locomotive - is very intense. It seems as there is an airborne part of the noise and a structural part, the latter vibrating through walls and other structures in the house. So every night you would hear and feel this subtle deep noise that made sleeping difficult. No ear-plugs were able to shut off that type of noise. The railways responded to a request for explanation that these old locomotives had to be warm to be able to start in the morning thus the need for them running idle all night. This environmental noise went on for 3 months a year.

The dogs: Your family home was a small townhouse with a front and backyard and with neighbors to both sides. To one side they had two dogs that were left alone from early morning when they went to work to late afternoon when they returned. Perhaps the dogs actually didn't like being alone: They barked all the time from around 07.00 – 19.00 in both the front and backyard that they could access through the garage. The noise from a barking dog can be very loud and it is very sudden. Perhaps the human response to the noise is some sort of "fear" response: You get a little shock each time. And it was very difficult to get used to this type of noise. If you tried to work in your home office it was nearly impossible to think due to the barking noise from the dogs. Then you tried to move around in the house but the dogs were still there and it seemed impossible to find a noise free place in the house. After a while you finally decided to confront the neighbors with the problem but as they "do no see the problem" they do nothing. To be able to do your work (a lot of concentrated reading and writing) you invested in a set of noise-cancelling headphones; what a relief and what a cool invention that made working at home possible in spite of the dogs barking.

Air travel: This example of how noise bothers you is not from one single air travel; it is from all your air travels. Every time you board an airplane you are bothered by the intense and constant noise in the cabin. In fact, according to a recent study[7], the steady noise during the flight is around 85 dB. When you sit in this constant and low frequency noise for many hours, on long flights up to 8-12 hours, it has a very "battery-draining" effect on you. Again this noise is both air-borne and is reflected through the materials in the cabin such as seats, arm rests and tables. You may have tried with high performance earplugs (>30 dB) perhaps without any significant improvement. The best solution for you has been the noise eliminating headphones. One of the advantages of using these headphones in an airplane is that you can still hear if there are announcements or if the person next to you says something. And then they strongly reduce the airborne noise to an acceptable level. The only problem with this over the ear headphones is that they are difficult to sleep with (and you need to sleep if you wish to work when landing). So not a perfect solution but definitely less "battery-drained" when you arrive to the destination.

A new road: Heavier traffic demands more roads. Near your home they approved building a new road through terrain that previously consisted of green fields. No news here – this happens all the time all over the world. A large group of experts estimated the potential impact of building this road on the environment including the potential for sound pollution. Lengthy formulas assuming number of cars, trucks and other vehicles x level of noise in dB x speed on the road and taking into consideration the distance to the nearest condominiums. This all led to the conclusion that "all seemed fine" with no increased noise pollution. When the road was built it was a nearly 4 km straight road with a perfect surface for riding motorbikes at high speed (something the experts did not foresee). Normal traffic of cars, busses and trucks produced some noise but at a distance of 300 meters this was not a problem. However, motorbikes that had been tuned to racing bikes and where both the engines and the exhaust pipes had been modified gathered from all over the area to race. Such motorbikes can reach an exhaust noise over 105-120 dB or similar to the take off of a jet airplane. The result is a noise that is overwhelmingly loud at all times of day and night. The noise emitted from these modified motorbikes was like 2-3 airplanes taking off at the same time. Ear plugs during sleep was the only weapon against this type of noise that could still be heard even with the ear plugs.

Open office space: You worked for "cutting edge" companies that put a lot of investment into building state-of-the-art office environments for their employees. Human resources studied the latest research on how could groups in the different departments work better together. As the company was successful it obviously expanded leading to employees often being had to move to new and larger office space. You had to leave your own quiet closed office for the newest type of offices: The *open office space* - free of dividing walls and designed to foster better team work. Suddenly you were together with 40 other colleagues in one big room; this room had a kitchen facility where colleagues could meet and have a cup of coffee and a glass room for work related meetings. This changed your work day with new and constant noise. With a positive spirit to the changes chosen by the company you did give this open office space a chance. The experts told you that you after some time would get used to the change and that you would then see increased collaboration. After nearly a year you only experienced disturbing noise (that you could not prevent or flee from) that negatively impacted your work. Again the best solution for you was to use the noise reducing earphones that enabled you to focus on the reading/writing that was part of your work. However, intellectual discussions were difficult and the open space noise interfered with any speech communication you had with work colleagues.

The snap-shots above are only a few examples of typical daily situations where noise from the environment could disturb your life. Daily environmental noise can impact our sleep, work or life quality while being at home. Perhaps you have had similar experiences as noise from the environment is everywhere and difficult to avoid. But is this noise really bad for us or our lives? The next Chapter will take a look at the potential health impact of noise.

CHAPTER 3. SOURCES OF NOISE

If you had to list some of the main sources of noise in your own life it would probably be something like:

- Traffic
- Construction
- Machines
- Public spaces
- Houses
- Workplace
- Travel
- Music
- Pets

This is not a prioritized list of noise sources but it could be relevant to go through this list to narrow down some of the problems that arises from each source. Take a look and see if you recognize some of these noise sources from your own experience:

Traffic: Cars, motorbikes, trucks, trains and airplanes. Engine noise, rolling noise from the wheels, use of horns, braking noise from trains etc. One major and global problem. Traffic affects now all parts of the country and all parts of the world. The density of traffic in the cities has just increased over the years and so has the noise. The noise from traffic does not respect the time of day. Roads are spreading with the increased need for mobility. To be able to drive from point A to point B in the fastest possible way, roads are drilled into mountains, through mountain valleys, under and over water. Thus resulting in an increase in traffic density and hence increased noise. As nice the progress of road building through a mountain valley is for the traveler as sad it is when you sit on the mountain and listen to the roar from the vehicles echoing in the valley below. Old cars, trucks, trains and airplanes make more noise than newer ones. Regardless, they are all still allowed to be used.

Noise from motorbikes is regulated in many countries. This means that the motorbike cannot emit more than a certain noise from the exhaust (e.g., 70 dB at 5000 rpm) when rolling out of the factory. Motorbike owners then replace the exhaust pipes with ones that produces much more noise (say 110 dB). Why is it necessary: "It sounds great"!

Brakes on cargo trains make a lot of noise. So these trains are not used during day time – but at night where they produce an incredible noise at the most sensitive time of day – when you sleep.

Some airports allow night flights. So anyone living near them will be able to hear the noise from arrival and take off of airplanes.

Traffic has been identified as the one of the sources of noise leading to most annoyance to the citizens[8].

Traffic noise needs immediate attention by legislators

Construction: Construction of houses, roads, pavements, terrain etc. uses heavy machinery that emits noise. When planning large construction work in cities sometimes the emitted noise comes as a surprise to the planners. This was the case when the "Metro-ring" (underground subway) was being constructed under the city of Copenhagen in Denmark. The 30.000 residents living near the construction sites were affected by elevated noise levels that in some cases were 24 hours a day. Noise lasting for years.

In most major cities today you will see maintenance and repair construction of old buildings along with construction of new buildings and terrain. Often no or little attention is given to the potential noise the use of jack hammers and other heavy machinery emits. This noise is constantly affecting thousands of residents living near construction sites. When one construction site is finalized another one opens.

Another type of "construction" noise is the one from the many do-it-yourself people that use semi professional power tools to make their own work on their own homes. Power tools like drills, grinders, chain saws and lawn movers can easily produce noise above 100 dB. Normally this is done in the afternoon and evenings including during the weekends when you need to relax.

Construction noise should be considered by the "polluter", advertised and options should be given to residents

Machines: A lot of machines are used to make work easier. At 04.00 in the morning the garbage truck will come and empty the garbage bins. The huge lift will carry the bin to the "mouth" of the trucks container, empty it here and then compress the content. Each action (like emptying a bin) will emit a tremendous noise. About eight bins will be emptied twice a week. On other days a week the two bottle and glass containers will be emptied. A lift will carry the glass container over the truck and at a certain point drop the content of thousands of glass bottles into the truck. The sound produced here is overwhelming.

Streets are cleaned using motor driven street sweepers. Older sweepers tend to emit an incredible noise. In some cities street sweeping is done in the evening or at night due to minimal traffic. The result is an intense noise a bit like an airplane passing by the street.

Hydraulic pumps are used to pump liquids including hot water. They are used in industries and near housing buildings. Most of them work on demand when there is a need for applying a certain pressure to a system of liquids. They work day and night and emits an often very strong noise that can be heard far from where the pump is situated.

Agricultural machines needs a specific mentioning here. Tractors, harvesters and many more agricultural machines are important for our farmers in order to effectively conduct their daily work. Such agricultural machinery often work for many hours a day, sometimes up to 12-15 hours. These machines produce a significant noise with measured levels up to more than 90 dB with some specific machinery such as farm sprayers generating more than 100 dB. New machines make less noise when compared to old models[9].

Due to the very long exposure times of the farmers these machines pose a specific and direct risk to their health and hearing[9]. As these agricultural machines do not always stay on the fields or farms but are also moved via the normal roads through inhabited areas, they can pose a huge noise burden to normal citizens in their homes. It is hard to really understand why such agricultural machines cannot be produced similar to a car with sound reducing technology to protect the farm worker and the environment from noise.

Machines in general should be regulated in order to reduce sound emission

Public spaces: Anyone can go there; parks, squares, walking streets etc. You often go there to relax, sit on a bench, enjoy a coffee and read a newspaper. But public spaces are sometimes also having problems with too much noise. One bad habit is when loudspeakers are used in a public square or park to constantly play music. In some parks they have done such a good job placing them all over it that you are not able to find a quiet zone.

Street performance is exciting and fun to watch. The kids love it. However, when street performance ends up with constant music and other types of "artistic" noise it impacts the quality of enjoying the public spaces; it removes the quiet zones. Public spaces are often used for festivals and public performances. Such festivals can result in days of music, performance and other noise generating activities. When these festivals are done in densely populated zones many citizens are suffering due to the generated noise that they have no way avoiding.

All public spaces should have quiet zones

Houses: The way houses were built perhaps 30-40 years ago did not take noise emission or transmission much into consideration. A lot of houses today suffer from not being adequately sound insulated. The consequence is obvious: Within one home sounds are transmitted from room to room. Talk, music, TV, washing machine, tumble dryer, vacuum cleaner, drill and power tools – all can easily disturb the person in the next room. In apartments, sound can travel through floors, via air vents, through water pipes, along roofs and sometimes noise generated can disturb people even several floors from the noise source.

Modern homes are often built to reduce energy consumption. These constructions use materials that provides high insulation from heat or cold but at the same time also high sound protection.

When noise insulation also is a priority in house building the effect is really remarkable. Soundproof windows, flooring, walls, pipes, roof, ventilation channels can all dramatically reduce noise emission. If you need to play the piano or your drums, special music room design can basically eliminate any noise emission due to you playing the instruments.

Sound proofing should be a standard for new houses and should be incentivized for updating old houses

Workplace: In any work environment there is often a need for places where you can talk and socialize and places where you can relax and think. Noise in the workplace is a reality and the source can come from machines (like office machines) or from your co-workers (talk, music etc.). How the workplace is designed will be important for the noise sensitive workers. Are there any quiet rooms where they will be able to do work that requires some silence? As with normal houses, the choice of building materials is critical and can make an incredible difference; the same office before and after a reconstruction with an upgrade to noise dampening materials (roof, walls, floor, separating walls) was nearly unrecognizable. Noise from talking colleagues and office machines nearly disappeared due to the new materials used.

A workplace needs to be constructed in a way that allows quiet spaces so that the needs of noise sensitive employees is met

Travel: Travel in cars, trains and airplanes can be a noisy affair. The noise sources are multiple from people, material and machines that you encounter during the travel. Long travels are especially challenging and the accumulated noise really drains your batteries and makes you very tired. Long train rides begins with noisy waiting halls. No quiet zones here. In the train cabin you may be lucky if you can find a quiet zone; meaning a zone where talk and use of smartphone is prohibited. But still there is the train and track noise to deal with. If you have to change train you will encounter another waiting room in the station with all sorts of noise. No place to relax. The same with long air plane rides. Here the noise is very loud and constant. In airports you may see noise sensitive passengers desperately trying to find a small spot in the airport without calls, sounds, talk, music and all the other noise sources. Some airports offers quiet spaces in the business lounge – but that is only for the "selected" few. For both train and air cabins new technology in design and build makes a lot of difference with regards to reducing the cabin noise.

Travel providers need to ensure that travel can be made in a quiet way from start until final destination.

Music: Music is great in all its different genres. Music can stimulate kids to learn better and can bring old and young together to enjoy something together. But the concept of having music everywhere at all times is not productive. If you enter a shopping mall the music begins in the parking area. It will follow you on all the "streets" to the shops and each shop will play its own genre of music – loud – very loud. Similar in a hotel; lobby, bathroom, elevator, corridor, restaurant – and when you enter the room just out of courtesy they have switched the TV on with some welcome music for you.

Amusement parks are examples of the maximum music and noise experience where you are surrounded by music that is only amplified if you take a ride. It is difficult to find just one quiet spot here without music.

It could be that someone regards this type of constant music as relaxing for people – it might have just the opposite effect on others.

Music is also wearable now: You can listen to music anywhere you go. Your smartphone is loaded with digitalized music ready to be played when you wish and you can listen through your headphones, wireless speakers at home or in your car etc.

This mobility of music is a fact and can reach parts of the environment with sometimes a negative effect. Boom cars, i.e., cars with enormous built in stereo amplifiers and speakers, can produce >175 dB of sound. Such a car in a quiet park removes the sought after silence immediately. Boom boxes are small mobile wireless amplifier/speakers that can be transported walking, on bike or anywhere and can play the music on your digital device so everyone can hear it.

Perhaps some of you have became so used to music everywhere that the only way to relax is to listen to it. For the more noise sensitive part of the population this constant music is not relaxing at all.

Listening to music should be a personal choice; music in public places should be minimized.

Pets: Pets may be noisy but if they are it's the owner that's to blame – not the pet. The owner made the decision to have the pet and to raise it, he should educate it as well.

Dogs may bark with a loudness of more than 105 dB. Parrots and peacocks can produce screams of a whooping 115 dB. Cats can produce significant noise at night howling.

Being able to produce so much noise, pets are still being introduced to places where people live closely together such as in apartment buildings, town houses or houses where people live close to their neighbors. This noise can be loud and can occur day or night for many hours at a time.

Pets are a "sensitive" source of noise in the sense that the owners are sensitive about being approached with the issue. Nevertheless, this noise disturbs many citizens around the world.

Noise from pets is the owner's responsibility (not the pets). Regulators should take noise sensitive citizens into account to ensure that noise from pets is minimized to no more than a tolerable level.

CHAPTER 4. COULDN'T-CARE-LESS

One type of noise that deserves its own chapter is the one deriving from the couldn't-care-less-attitude of some people. This attitude leads to other environmental problems such as pollution of the air, water or soil. In these cases people near us do not take into account the existence of others. It just doesn't matter – they "couldn't-care-less":

> The neighbor playing loud music or listening to surround sound TV at night or using the chainsaw Sunday morning.

> The young man buying a high power motor bike and changing the exhaust pipe to an illegal one so now the bike has a loudness of 110 dB – then he drives through the city center at night.

> The person living in a condominium who acquires two new German Shepherd dogs – and then leaves them alone in the apartment 9 hours a day while he is at work.

> Parents that allow their kids to behave in a completely uncontrolled way thus

generating noise in hotels, public spaces, restaurants and perhaps most importantly in educational institutions such as schools.

The list is unfortunately much (much) longer. This is a problem that WHO, governments and international institutions do not cover that much. Perhaps because it is very difficult to control. It is about what people think (or perhaps rather do not think). It is nevertheless an important problem.

Once certain smokers had a similar attitude. Their "addiction" to tobacco was stronger than their will to take into account the people next to them. They smoked even in restaurants next to eating non-smokers. And when confronted with their attitude towards non-smokers, they often became very aggressive. Similar to the couldn't-care-less noise source: If you confront them, they often get angry; "it is my right to listen to music when I feel like it in my own home". But the smoker doesn't smoke anymore on air planes. No-one is really talking about this anymore. There has been a complete paradigm shift in how smokers accept the new (and improved) situation of no smoking on air-planes (or in restaurants or public buildings or in trains etc., etc.). And many people are today happy about the clean air in pubs, bars and restaurants, even the smokers….

So it can be done and perhaps we need to begin with education (already in school) on noise impact, changes in rules for when and where noise is acceptable (even in condominiums), laws regulating also "couldn't-care-less noise" and severe fines for violating the same laws.

Education on why noise is harmful to us human beings and why silence is positive and good is important. Perhaps many couldn't-care-less noise violators never have had the chance of enjoying silence. Some kids begin their institutional life in noisy kindergartens, then noisy schools and then noisy work places. Perhaps nowhere on this (noisy) route has this person had the opportunity to walk in a silent forest or on top of a silent mountain. Silence has never been part of her/his upbringing.

"Couldn't-care-less" environmental noise should theoretically be easy to fix – it is "just" a change in attitude that is necessary. However, in reality is often very difficult to change such an attitude and as such reduce the noise from this source.

CHAPTER 5. NOISE BURDEN AND HEALTH IMPACT

Put simply:

Noise affects a lot of people

- it is a burden on many citizens

- and noise kills.

The main global expertise on both the burden and the health aspects can be found at the WHO that have published multiple excellent and highly recommended publications and guidelines on environmental noise[3, 4, 10, 11].

WHO defines health as[12]:

> *Health is a state of complete physical, mental and social well-being and not merely the absence of disease or infirmity.*

From these expert sources we understand that environmental noise is a global problem that affects many people daily and significantly affects their health.

If we take European Union of app. 500 million inhabitants as an example, the burden of environmental noise (exposure of constantly > 55 dB) is substantial[5]:

- 125 million by traffic noise
- 8 million by railway noise
- 3 million by aircraft noise
- 300.000 by industrial noise

And these numbers do not include other sources of environmental noise such as domestic (neighbors), construction, tools, music, festivals etc.

In a recent study ordered by the EU commission on health impacts of noise[8], it was estimated that exposure to noise in Europe led to:

- 10.000 premature deaths due to heart disease and stroke
- 43.000 hospital admissions per year
- 910.000 additional cases of hypertension

WHO did a study report[13] on the burden of disease from environmental noise in Europe where they looked at the relationship between environmental noise and specific health effects such as cardiovascular disease, cognitive impairment, sleep disturbance, and ringing in the ears (tinnitus). The burden of disease was measured in terms of disability-adjusted life-years (DALYs: One DALY can be thought of as one lost year of "healthy" life). Based on this WHO estimated that DALYs lost from environmental noise were:

- 61 000 years for ischemic heart disease
- 45 000 years for cognitive impairment of children
- 903 000 years for sleep disturbance
- 22 000 years for tinnitus
- 654 000 years for annoyance

Summarizing this:

- **At least one million healthy life years are lost every year from traffic related noise in the western part of Europe.**

And again, the health impact was only evaluated on noise generated from traffic sources. So it is likely that many more people may have health implications due to noise.

Numbers from other places of the world like the US are comparable to EU defining noise as a major burden and a major health problem[14].

Noise is a huge problem globally and has tremendous health implications to citizens around the world.

Let us take a closer look at the science behind the health impact of noise:

Heart disease and stroke

WHO concludes[15]: *There is sufficient evidence that long-term exposure to community noise increases the risk of cardiovascular diseases.*

There are several paths to which noise can impact heart disease and stroke. Noise may affect changes in circulation, including blood pressure, heart rate, electrocardiogram, cardiac output and vasoconstriction as well as the release of stress hormones (catecholamines, adrenaline, nor-adrenaline, and cortisol)[15]. Elevated blood pressure, blood lipids and blood clotting factors are seen in people exposed to high levels of noise.

Increase in noise levels from transportation is linked to increase in cardiovascular disease; for every 10 dB increase in noise level the risk of cardiovascular disease increased by 7-10%[16].

Specific increased risk of stroke due traffic noise was evaluated in a recent literature review (meta-analysis) where all literature on traffic noise and stroke was searched; an increased risk of stroke (> 5 %) was found in people exposed to mainly air traffic noise[17].

Sleep disturbance

Sleep has, from a medical and psychological point of view, a profound impact on our well being and general health. A good sleep, every night, is important for how well we study, work, interact and live. It is becoming an area of research to better understand how sleep deprivation may potentially lead to diseases such as cardiovascular disease, obesity and diabetes, psychiatric illness and cancer[18].

WHO evaluated the available knowledge on the impact of noise on our sleep in a recent report that concluded[10]:

> *There is sufficient evidence that night noise is related to self-reported sleep disturbance, use of pharmaceuticals, self-reported health problems and insomnia-like symptoms. These effects can lead to a considerable burden of disease in the population. For other effects (hypertension, myocardial infarctions, depression and others), limited evidence was found.*

These guidelines suggest night time levels of noise not to exceed 40 dB ($L_{night, outside}$)[i] to protect the public.

---

[i] Refers to the EU definition in Directive 2002/49/EC: See details in Glossary.

There is reason to believe that the impact of noise on sleep and the resulting health issues will be the focus of much future research[18].

Learning problems

We learn during our entire life. In our childhood and early adulthood we go to school and perhaps later university or work place.

In all phases of life it is important that we continuously study and learn. As noise may impact our sleep, one of the results could be how well we learn the day after just one single night of bad sleep. Sleep disturbances, such as noise, leading to poor quality sleep may have an impact on the brain functions and how well we learn.

In a literature study looking at young and older adults, it was concluded that sleep quality is important for the learning ability in younger adults and middle aged people and that good sleep may even protect for loss of for e.g. memory later in life due to advanced age[19].

- **Good sleep is linked to good learning**

For children noise in the daily environment such as in her/his school has also an impact on learning. In an interesting overview on "noise and children's health", there was a difference in how well the children performed in school depending whether they had a "noisy" or "quiet" school environment; a quiet school environment resulted in better school performance[20].

In healthy adults noise in the daily environment also negatively impact the learning abilities; in a study, healthy adults exposed to urban noise or social noise performed worse in learning tasks than when exposed to a quiet surrounding[21].

New studies on pre-term infants cast light on the impact of noise on tactile learning; a noisy environment is not as good as a silent environment when it comes to learning for these small neurologically not mature beings[22].

- We all learn better in a noise free environment

Mental health

WHO defines mental health as[23]:

> "*a state of well-being in which every individual realizes his or her own potential, can cope with the normal stresses of life, can work productively and fruitfully, and is able to make a contribution to her or his community*"

Environmental noise has the ability to impact how we sleep, how we learn, how we work, and in general how we feel. Our mental health is defined by both genetic factors and by our environment. In a literature study evaluating the noise effects on mental health in adults, ambient noise was associated to not only an impact on learning, memory capacity and activities of daily living but also to depressive symptoms, elevated anxiety and nuisance[24]. This could mean that there is some kind of association of environmental factors with mental health, but a firm conclusion is missing and will require future research[24].

Recently a group from Germany studied the relationship of the three main sources of traffic noise (road, rail and air) on depression, one of the major mental health diseases. The study was large and the findings disturbing but very interesting; for all three noise sources a statistically increased exposure-risk relationship between noise and depression was identified (the increased risk was from 10-23 %)[25]. The conclusion was that traffic noise exposure could actually lead to depression.

Overall impact of noise

It is clear that noise affects and annoys a lot of people of any gender, any age and without any boundaries (ethnic, social, etc.). In many ways you could say that:

Noise is a silent killer

It sneaks in on you and slowly disturbs your sleep, concentration, work performance, social activities and your quality of life. Then you may even become sick as described above without recognizing that you are ill, all because of noise from your environment.

So having come to this understanding, it is clear that we need to do something about this "silent killer". In the next chapter we will take a look at what is being done about the environment noise.

CHAPTER 6. INITIATIVES TO LIMIT NOISE

All over the world politicians and governments recognize that noise is a global problem. As noise is a global health problem, it is only natural that the WHO is spearheading research on the global noise issue and its effect on the citizens of the world[4].

The activities by the WHO are really noteworthy and very extensive ranging from noise research on health impact assessment to publication updates of noise guidelines with the latest version dating back to 1999[3]. A revision and update of these guidelines is awaited. These guidelines are indeed important as they should be driving global political efforts on setting limits for what should be acceptable noise to avoid harm to the citizens of the world.

And legal noise directives are indeed being written; one example is the EU environmental noise directive from 2002[26] that has a lot of relevant recommendations and guidelines to acceptable levels and activities in the European Union on how to bring noise levels down. Similar guidelines and laws can be found in Japan[27], in the US[6] and many other countries all over the world.

That is the good news.

But there is more good news: Politicians and governments and experts conduct a lot of meetings and write a lot of expert reports on environmental noise where they address the important issue of the damaging impact of this noise. Here they discuss how to:

- Conduct relevant research on noise
- Predefine threshold values for noise
- Perform noise mapping
- Define priorities on what to do within relevant budgets
- Outline regulations and directives on noise
- Define means to minimize noise

However, there is also some bad news to the many (read: incredibly many) people that are affected by noise in their daily life: In spite of a general acceptance by politicians and governments that noise is bad and in spite of all the relevant attempts to reduce noise levels, the noise threat is just becoming greater.

On April 24, 2017 the EU Commission conducted a conference "Noise in Europe" that can be seen on a webcast[28]:

Both attending politicians and experts agreed on the fact that environmental noise (from traffic) is one of the major environmental challenges in the EU region.

The panel was asked a relevant question:

"*Has the EU noise legislation failed?*"

The politician from Finland had the most relevant and telling answer:

"*As a politician I am disappointed in myself. Today we are living (in) 2017 and there are still 16.000 premature deaths (due to noise). This is enough for an answer. It is too much so the steps have been too small*".

This is an honest and relevant feedback to us, the citizens of a world where noise is a growing problem despite regulations, research, noise mapping and noise guidelines. Whatever impact these initiatives may have had on noise, perhaps it is fair to say it has been slow – very slow – and not fast enough.

So the politicians are generating goals, guidelines and laws for reducing environmental noise; and despite this important effort the goals are not met.

Norms for noise reduction of new cars, trains and airplanes was discussed as one way forward. But just to highlight the complexity and difficulty our politicians are facing, let us take trains as an example with regards to introducing these noise norms on new train models: If the old trains are allowed to continue on the tracks – the average life time of a train wagon was mentioned to be 40 years before they are phased out – so it could take up to 40 years for the noise sensitive citizens before they feel any difference of a noise reducing initiative. That is just not acceptable!

- **Noise legislation is not having the intended impact as noise continues to grow as an environmental problem**

One of the issues that was discussed in this EU noise conference was the impact of noise on social inequality; if you can afford it, you can buy a life with silence – in other words:

Noise affects social classes differently.

One example from the conference was interesting: Noise from roads or train tracks or airports could devaluate the price of houses to levels that were affordable to low income groups of citizens. This devaluated house value would then create social inequality; only people with the low income could afford these homes and would be forced to accept higher exposure to noise with all the health implications related. But the social inequality due to noise is not only limited to housing; this can be found where noise impacts institutions, educational system like schools, work place etc.

The right to silence should not depend on your social class.

So something is not working here. Guidelines on noise are there. Noise is being mapped and research is begin conducted. Noise is a known burden and a health risk. Laws have been made to reduce noise. Politicians and governments are "talking the talk".

But very little is happening.......

Chapter 7. Ways to seek silence in daily life

In this chapter we will have a look at different ways on how to seek silence or at least minimize noise from your environment. Seeking silence in our daily life begins with our own way of living; how do you generate or avoid generating noise and how do you ensure that you maintain a silent environment?

Be sensible to your own noise: Before pointing at all the other sources of noise in our environment, we need to begin with ourselves. We must try to be sensible to the noise we are generating and try to avoid it as much as possible. We should remind our family members to be sensible too; avoid playing loud music, do not shout in the house, be careful with the TV volume, do not take the elevator at night etc.

Noise vigilance: It is important to dare to be vigilant to environmental noise. As it seems as if the different politicians are aware that noise is an annoyance and a health problem, perhaps not all citizens are aware yet. As such, you need to show some noise vigilance and identify to the competent authorities if environmental noise exceeds the levels of (in this case the only way you can measure this) your personal threshold.

If noise becomes annoying, contact the neighbor or the owners association where you live. Of course in a kind and factual way.

The mayor or the municipality could be another place to contact if the noise source is within their jurisdiction. Sometimes the police needs to be involved depending on the source of the environmental noise.

By being vigilant you participate in setting better standards for unacceptable environmental noise levels. By not doing anything – noise levels will only increase.

Teach our children about silence: We have a lot to teach our children with regards to reducing the noise we generate and seek the beauty of silence. Our kids need to experience the sound of the sea without music from their cell phone. It is great to take them to the mountains perhaps filled with snow where there is absolute silence and just sit and enjoy it with them. If you go camping in a tent they might wake up the first morning with the sound of the birds – sounds they have most likely never heard before. If our kids never experience silence perhaps it is one of the reasons they will not appreciate it when they grow older. Nor will they actively seek it.

Plan silence: During your workday you can be exposed to a lot of noise. Or if you travel, you know already that the noise exposure will make you very tired. Knowing this you may plan for moments of silence. You could choose lunch in a quiet place or immediately after work a walk in a quiet park. During a long travel you may need to find spaces that are without noise (not always so easy). Your cell phone needs to be recharged – and so do you: Seeking silence is your way of re-charging.

In your spare time you could plan for silence. One of the ways is kind of sport activities that include silence; like fishing where you can seek places in the nature with small streams where the nature is beautiful and there is absolutely no environmental noise. With your partner you could go Nordic walking in the nearby mountains. Small tracks away from any modern noise generating source and away from roads. You may be walking together often without talking just enjoying the beauty and the silence of the isolated mountains.

Identify a silent home: The first step here is to identify a home that from the beginning is silent. That is not easy! There are many things to consider such as: How the home was built (use of noise reducing materials)? What noise sources surround it (traffic, construction, airports etc.)? Do your neighbors share your values for enjoying silence? All extremely difficult to evaluate before moving into a new home to avoid any negative noise surprises. Today it is possible to use Google maps to take a look at what kind of noise sources are detected near a potential new home. Also, before signing a contract you may find it valuable to stay a day or so in a house to understand how well it is sound insulated. Windows, walls, doors, floors all transmit sound and all can be built using materials of different sound proofing quality. Material from before 1980 is most likely not built according to any modern sound proofing standard.

One good example is modern soundproof windows that can in no way be compared with old window technologies. They are so good (and so expensive) that buildings close to a railway can prevent most of the train noise after their installation. If the existing materials in your new home are not up to current sound standards, you may consider updating them to gain a much quieter home. Alternatively, you may look for modern homes built specifically with sound proof materials. The results are really noticeable but again this "luxury" comes with an increased cost.

Block the noise: Blocking the noise is not a general solution to the environmental noise – reducing the noise is. However, the one thing that we always should be carrying is ear plugs; to block any unwanted noise especially during sleep. These can be found as disposable polyurethane foam earplugs that are soft self expanding to fit your outer ear canal. There are many ear plug qualities and you need to research which of these has the best Noise Reduction Rating (NRR). Some of the best have a NRR of about 34 dB that is enough to reduce noise from traffic or from your spouse snoring (another significant reason for disturbed sleep). You could use ear plugs on especially travel as they are comfortable in an airplane where they allow you to rest your head in any position if you need to sleep (unlike sound cancelling headphones that prevent you from resting on your side). However, if used daily such ear plugs may have adverse effect in your outer ear canals such as irritation or even infection.

Earplugs (with a high NRR) offer also important protection of your hearing in situations with extremely high noise such as while hunting (gun shots), work with loud machines, rock concerts etc. Such high sounds can provoke high-frequency permanent hearing loss. This is better avoided with a couple of good ear plugs.

As mentioned earlier another way to block noise is sound cancelling headphones. There is no need to advertise for a specific brand here, as they are generally way too expensive. That said, one brand really stands above the others. They produce an over the ear headphone where there is a built in microphone system that detects the characteristics of the noise from the environment to generate a counter noise within the headphone: The effect is excellent as it may give you hours of near silence. You can use these light headphones while working and on very long air travels during your time awake in the air cabin. You will have to do a little research to identify what headphone type you like and most of all you will need to try them on to see what fits your ears the best. Airport shops or malls are good places to try such noise cancelling headphones – here you always find a lot of noise and the noise cancelling capability is immediate.

Chapter 8. Goals for noise reduction

Smoking kills – we have all accepted that

Noise kills – we are not there yet

The campaign against in-flight smoking, i.e., smoking while in an air cabin is an example of how effectively an environmental pollutant has been eliminated.

In the 60's and 70's it was normal to sit in air cabins full of smoke. It really did not matter if there were sections for smokers and separate sections for non smokers; all were breathing the same air. Consider having a 10-hour flight where the air was thick of smoke.

In the US as an example, the US surgeon general's report on Smoking and Health in 1964 (only) suggested a link between smoking and cancer. In 2006 came a report on the health impact on passive smoking. These reports, supported with a lot of clinical evidence, woke up a movement against smoke in public places in general. Especially in-flight smoking was limited over 3-steps; in 1988 with a smoking ban on all domestic fights of less than 2 hours; extended in 1990 to include domestic flights > 6 hours and in 2000 to include all international and domestic flights. Thus a total smoking ban over 12 years: No smoke on airplanes.

Today it is hard to imagine how it was entering a flight cabin full of smoke – we are so used to a smoke free environment not only in air planes but in trains, public buildings, restaurants, hotels, pubs etc. And violating these rules and laws has immediate consequences for the "polluter": If you smoke in a US plane you may be fined up to 5000$ and even arrested upon arrival or worse be responsible for a pre-mature landing of the air plane. The point with this example is that history shows that it can be done. We now need responsible politicians to take action and ensure that noise is banned or reduced to levels acceptable for our citizens. For all our citizens regardless of social status because like smoke kills – also noise kills.

It is not clear what was the "the last drop that made the cup run over" in the example of the in-flight smoking: What made the airline companies ban smoking? Was it common sense in avoiding exposing a lot of citizens to unhealthy smoke or fear of law suits from people later getting sick of passive and involuntary smoking? Whatever it was, we need to learn from it in order to make a radical change in environmental noise.

So environmental noise needs to be minimized immediately – the longer we wait the more people will suffer. In this chapter we will try to focus on some stringent goals for how this could be done as soon as possible. Time is critical.

So we need to do something and it has to involve governments throughout the entire planet! However, as mentioned in Chapter 5, in spite of the very clear messages from the WHO that noise disturbs a lot of people and make them sick, and in spite of guidelines (WHO guidelines on environmental noise are from 1999![3]) on noise limits and directives and laws by governments, the problem is just getting worse. A lot of research, meetings, reports, discussions have been conducted – it is now time to act on these.

Less talking – more walking.

Because, to use the analogy with in-flight smoking: There is "still smoke in the cabin". To get to a tangible result we need first to define some simple goals to reach the vision:

A society where silence is prioritized over noise.

To make such changes it will be important that governments, countries, regions, and united global organizations work together with the common understanding that:

1. It has to be done – and quickly
2. Change will cost money and in some cases impact businesses
3. Global leaders need to be held accountable for reaching these goals

Below are some basic goals that perhaps could get us there:

- Stepwise and achievable guidelines: Generate simple and achievable guidelines within a reasonable time frame (read: Reasonable for the population now being "noise hostages").
- Implement guidelines in new laws: Each country needs to generate new and simple laws and enforce them following the stepwise guidelines.
- Enforce laws: Polluter pays principle: As with any other environmental polluter, the person/organization responsible for noise pollution has to pay. Deliberate and careless noise pollution should never pay off.

- <u>Fines and taxes on noise pollution</u>: Heavy fines should be given to the noise polluters. In order to speed up the process of achieving the goals taxes on noise pollution (e.g., on older noisier machinery) should be introduced.
- <u>Noise mapping app</u>: Qualitative noise mapping should be introduced through a global app where people can signal annoyance to a noise source. Areas of unacceptable noise should trigger intervention by authorities.
- <u>Sound marking of all homes:</u> Similar systems as energy efficiency marking of all homes where you get an idea of how noise intense a home is. This marking can be centralized so that politicians can invest in areas that need to be noise reduced.
- <u>Source noise reduction incentives</u>: It should pay off to reduce noise at the source level and as such incentives are needed.
- <u>Noise shield incentives</u>: Improved sound proofing of old houses, eco-friendly sound barriers and other noise shields should be heavily incentivized.

It is obvious that the suggested goals here are big and that achieving them will involve a huge collaboration between politicians, governments, institutions, law enforcement and every single one of us citizens. But we need to do it. We need to elect politicians that understand the burden and problems that environmental noise puts on the citizens.

As citizens we might have to sacrifice a bit of the speed of the technological development in order to finance the extra costs of reducing noise. It may take a bit longer getting new technologies but it will all be more sustainable with the added benefit of better quality of life due to more silence and less noise.

CHAPTER 9. NOISE REDUCTION INITIATIVES

To stimulate some creativity here are some smaller and larger examples from the world on what is currently being done to tackle the growing noise problem towards a more silent ambiance.

Night noise guards: The Municipality of Copenhagen, Denmark, has, as with many modern cities, had problems with noise in the city center due to nightlife; bars, restaurants, music places etc. have for years generated a significant noise and prevented a lot of citizens from sleeping despite laws and rules against such noise. To minimize this problem, citizens who are disturbed from 21.00 – 03.00 can call a central that will trigger that a noise guard is sent to the polluter with noise measuring equipment[29]. The guards collaborate with the police and in case of violation they can immediately close down the establishment.

Without laws, vigilance, and adherence to the laws, noise will win over silence.

Quieter home initiative: This active initiative by Heathrow Airport, London, UK, included app. 1200 homes located near the airport who experienced the highest level of aircraft noise[30]. Home owners had acoustic tests performed in their houses followed by an evaluation of the potential improvements that would lead to noise reduction such as loft and ceiling insulation, sound proofing of windows and doors etc. If eligible for this scheme, the home owner would have all costs covered for the recommended sound proofing upgrades.

Government and politicians need to map areas of increased noise and need to be actively searching for citizens that potentially suffer from environmental noise and incentivize noise reduction activities.

Noise Management Incentive Scheme: Citizens of the city of Adelaide in Australia have the possibility of having a sound engineer evaluate the home for noise issues and to come up with recommendations on sound improvements for their house. These improvements are incentivized so that a percentage of the expenses are covered by the City[31].

Noise should effectively be reduced following guidelines and state-of-the-art standards and the citizens should be financially incentivized to do so.

Noise reducing roads: In Japan a lot of research has been conducted to evaluate new types of pavements[32]. The goals are to construct economical, durable and relevant to this book: Silent roads. By modifying the composition of the asphalt, particle size, binder and layer design a potential gain of 10 dB in sound reduction is claimed. With new sound guidelines for road noise, the use of new road pavements could potentially and immediately achieve the goals of road noise levels in Japan.

Traffic noise is one of the main sources of environmental noise and approaching the source of noise (the road pavement) makes a lot of sense.

Restricted Traffic Zones: In major cities traffic congestion is a well-known phenomenon. In the city centers, the excess amount of traffic has an impact on air and sound quality and thus the quality of living of the citizens. In Italy in cities like Rome and Firenze the so called Zona Traffico Limitato (controlled traffic zone) or ZTL has been introduced as zones where the traffic is severely restricted. This has an immediate and significant impact on the noise levels day and night. If you enter a ZTL zone with a vehicle the target plate will be automatically photographed. In case you have no authorization to enter this zone the consequence is immediate as you will be fined.

Where city centers are suffering from dense traffic and noise 24/7, restricted traffic zones may immediately have an effect – this obviously only works if the laws are being enforced.

Road toll on noise emission from vehicles:
Since 2017 on highways in Austria, large vehicles such as motor homes, busses and trucks have to pay a surcharge for traffic related noise pollution following the "polluter-pay" principle: The more you contribute to noise pollution the more you pay in road toll. Interestingly, between 22.00 and 05.00 higher night road toll rates apply to minimize noise at night[33].

Taxing the polluter may modify behavior. This principle could be adapted to other types of environmental noise polluters. It should not pay to pollute!

CHAPTER 10. A VISION FOR A QUIET FUTURE

A **vision** for our future is that everyone on the planet, regardless of where you live, your social status, and your educational background will be able to live together in:

A society where silence is prioritized over noise

The steps to achieve this vision are that everyone on the planet will (**mission**):

- Educate on the benefits of silence over noise
- Accept simple global guidelines on how to reduce noise
- Secure access to silence in all aspects of life
- Push lawmakers towards universal and better noise regulation – effective NOW
- Follow laws on noise regulation
- Agree with the enforcement of noise laws

With this vision and mission statement hopefully we will all be able to enjoy "*More silence less noise*" and that environmental noise will not disturb our living (anymore).

CHAPTER 11. NOISE RELATED WEB SITES

Noise is a global problem affecting citizens everywhere regardless of what parts of the world they live. Below are just a bit of multiple available web sites on this topic – just to get you started:

World Health Organization (Global). The global expertise in noise burden and the impact of noise on health. Really recommended with excellent reports and global guidelines on noise and health: www.euro.who.int/en/health-topics/environment-and-health/noise

EU Commission website on Noise (Europe). European Union guidelines and directives. Links to important event/webinars on environmental noise: ec.europa.eu/environment/noise/index_en.htm

The Right to Quiet Society (Canada). For Soundscape Awareness and Protection. Good links and resources to noise related matters with global affiliations: www.quiet.org

Noise Free America (US). An organization dedicated to fighting noise. Good resources to noise related issues focused on the US but with universal inspiration. Multiple global links: www.noisefree.org

Noise off (US). US focused noise resource with links and an impressive library to noise related reports: http://www.noiseoff.org/

The Norwegian Association Against Noise (Norway). Independent anti-noise organization: www.stoyforeningen.no

Missione Rumore (Italy). Associazione Italiana per la difesa dal rumore. Founded by citizens disturbed by noise. Expert advice and links to local cases of environmental noise: www.missionerumore.it

The UK Noise Association (UK). Voluntary campaign for action against noise. Local links to anti noise sites: www.ukna.org.uk

Awaaz Foundation (India). Environmental foundation against noise and other pollution. Activities and campaigns against increasing noise in India: www.awaaz.org

GLOSSARY

Term	Definition
Blood lipids	Blood fats circulating in the blood either free or bound to other molecules.
Cardiac output	Cardiac output is a term used in cardiac physiology that describes the volume of blood being pumped by the heart, in particular by the left or right ventricle, per unit time.
Electrocardiogram	Electrocardiography (ECG or EKG) is the process of recording the electrical activity of the heart over a period of time using electrodes placed on the skin. These electrodes detect the tiny electrical changes on the skin that arise from the heart muscle.
Hz, Hertz	Hertz is a measure of frequency; one cycle per second. 1 KHz is 1000 cycles per second.
$L_{Aeq,T}$	Exposure to noise for the duration of a given time interval T (a 24-hour period, a night, a day, an evening) is expressed as an equivalent sound pressure level (measured in dB(A)) over the interval in question.
L_{Amax}	Maximum outdoor sound pressure level associated with an individual noise event.
L_{night}	Refers to the EU Directive 2002/49/EC: Equivalent outdoor sound pressure

	level associated with a particular type of noise source during night-time (at least 8 hours), calculated over a period of a year.
Noise Reduction Rating (NRR)	Noise Reduction Rating (NRR) is a unit of measurement used to determine the effectiveness of hearing protection devices to decrease sound exposure within a given working environment.
SEL	Sound exposure level is an equivalent outdoor sound pressure level associated with an individual noise event, with the equivalent level standardized at one second.
Sound pressure level in dB	Sound pressure level (SPL) or acoustic pressure level is a logarithmic measure of the effective pressure of a sound relative to a reference value. Sound pressure level is measured in dB a unit for expressing the relative intensity of sounds on a scale from zero for the average least perceptible sound to about 130 for the average pain level.
Stress hormone	Commonly known as the fight or flight hormone, it is produced by the adrenal glands after receiving a message from the brain that a stressful situation has presented itself. Three typical stress hormones are: Adrenaline, nor-adrenaline, and cortisol.
Vasoconstriction	Narrowing of the blood vessels resulting from contraction of the muscular wall of the vessels.

ABOUT THE AUTHOR

Author's vision:

A society where silence is prioritized over noise

Peter Kruse, MD, PhD, is an independent clinical consultant based in Europe. Dr. Kruse has worked as a hospital physician and as a physician in the pharmaceutical industry.

Dr. Kruse has through his entire life preferred silence over noise. He has lived in Asia, Northern and Sothern Europe and in the USA. As such he has experienced some of the global differences with regards to environmental noise.

Dr. Kruse is not a medical expert in noise. He has no political or regulatory expertise in the field of noise nor has he any agenda for writing about environmental noise other than one: He is a concerned global citizen that sees environmental noise as one of the major threats to the wellbeing of the citizens of the world.

More Silence Less Noise at: www.facebook.com/More-Silence-Less-Noise-961475810622876 where you will also be able to post a message or ideas about noise.

REFERENCES

(1) Dictionary.com. Silence. Define Silence at Dictionary.com. Internet 2017;Available at: URL: http://www.dictionary.com/browse/silence.

(2) Kagge E. *Silence: In the age of noise.* 2017.

(3) Berglund B, Lindvall T, Schwela DH. Guidelines for community noise. WHO.: World Health Organization, Geneva; 1999.

(4) WHO. WHO: Website on noise. Internet 2017;Available at: URL: http://www.euro.who.int/en/health-topics/environment-and-health/noise.

(5) EU. EU Commission website on Noise. Internet 2017;Available at: URL: http://ec.europa.eu/environment/noise/index_en.htm.

(6) EPA. U.S. Environmental Protection Agency website. Internet 2017;Available at: URL: https://www.epa.gov/.

(7) Ozcan HK, Nemlioglu S. In-cabin noise levels during commercial aircraft flights. *Canadian Acoustics; Vol 34, No 4 (2006)* 2006.

(8) Houthuijs DJM. Health implication of road, railway and aircraft noise in the European Union.: National Institute for Public Health and the Environment, The Netherlands; 2014.

(9) Depczynski J, Franklin RC, Challinor K, Williams W, Fragar LJ. Farm noise emissions during common

agricultural activities. *J Agric Saf Health* 2005;11:325-34.

(10)   WHO. Night noise guidelines.: World Health Organization, Copenhagen; 2009.

(11)   Braubach M, Jacobs DE, Ormandy D. Environmental burden of disease associated with inadequate housing.: World Health Organization, Copenhagen; 2011.

(12)   WHO. Preamble to the Constitution of WHO as adopted by the International Health Conference, New York, 19 June - 22 July 1946; signed on 22 July 1946 by the representatives of 61 States. Official Records of WHO, no. 2, p. 100. 1948.

(13)   WHO. Burden of disease from environmental noise. Quantification of healthy life years lost in Europe. Internet 2011.

(14)   Hammer MS, Swinburn TK, Neitzel RL. Environmental noise pollution in the United States: developing an effective public health response. *Environ Health Perspect* 2014;122:115-9.

(15)   Hellmuth T, Classen T, Kim R, Kephalopoulos S. Methodological guidance for estimating the burden of disease from environmental noise. 2012. World Health Organization, Geneve.

(16)   Basner M, Babisch W, Davis A, Brink M, Clark C, Janssen S, Stansfeld S. Auditory and non-auditory effects of noise on health. *Lancet* 2014;383:1325-32.

(17)   Dzhambov AM, Dimitrova DD. Exposure-response relationship between traffic noise and the risk of stroke: a systematic review with meta-analysis. *Arh Hig Rada Toksikol* 2016;67:136-51.

(18) Laposky AD, Van Cauter E, Diez-Roux AV. Reducing Health Disparities: The Role of Sleep Deficiency and Sleep Disorders. *Sleep Med* 2016;18:3-6.

(19) Scullin MK, Bliwise DL. Sleep, Cognition, and Normal Aging: Integrating a Half-Century of Multidisciplinary Research. *Perspect Psychol Sci* 2015;10:97-137.

(20) Paunovic K. Noise and children's health: research in Central, Eastern and South-Eastern Europe and Newly Independent States. *Noise Health* 2013;15:32-41.

(21) Wright BA, Peters ER, Ettinger U, Kuipers E, Kumari V. Moderators of noise-induced cognitive change in healthy adults. *Noise Health* 2016;18:117-32.

(22) Lejeune F, Parra J, Berne-Audéoud F, Marcus L, Barisnikov K, Gentaz E, Debillon T. Sound Interferes with the Early Tactile Manual Abilities of Preterm Infants. *Sci Rep* 2016;6:23329.

(23) WHO. WHO mental health definition. Internet 2017;Available at: URL: http://www.who.int/features/factfiles/mental_health/en/.

(24) Tzivian L, Winkler A, Dlugaj M, Schikowski T, Vossoughi M, Fuks K, Weinmayr G, Hoffmann B. Effect of long-term outdoor air pollution and noise on cognitive and psychological functions in adults. *Int J Hyg Environ Health* 2015;218:1-11.

(25) Seidler A, Hegewald J, Seidler AL, Schubert M, Wagner M, Droge P, Haufe E, Schmitt J, Swart E, Zeeb H. Association between aircraft, road and railway traffic noise and depression in a large case-

control study based on secondary data. *Environ Res* 2017;152:263-71.

(26) EU. EU environmental noise directive. Directive 2002/49/EC. Internet 2017;Available at: URL: http://eur-lex.europa.eu/legal-content/EN/TXT/HTML/?uri=CELEX:32002L0049&from=en.

(27) Ministry of the Environment. Japanese Noise Regulation Law. Internet 2017;Available at: URL: http://www.env.go.jp/en/laws/air/noise/.

(28) EU Commission. Webinar: Noise in Europe. Brussels, Belgium. Internet 2017;Available at: URL: https://webcast.ec.europa.eu/noise-in-europe.

(29) Copenhagen Municipality. Copenhagen municipality noise guards. Internet 2017;Available at: URL: http://www.kk.dk/støj.

(30) Heathrow Airport. Heathrow Airport Quieter Home Initiative. Internet 2017;Available at: URL: http://www.heathrow.com/noise/what-you-can-do/apply-for-help/quieter-homes-initiative.

(31) City of Adelaide. Noise Management Incentive Scheme. Internet 2017;Available at: URL: http://www.cityofadelaide.com.au/your-council/funding/noise-management-incentives/.

(32) Danish Road Institute. Noise reducing pavements in Japan. Internet 2015;Available at: URL: http://www.vejdirektoratet.dk/DA/viden_og_data/publikationer/sider/publikation.aspx?pubid=000058673.

(33) Austrian Highways ASFINAG. Distance-related toll including surcharges for air and noise pollution. Internet 2017;Available at: URL: https://www.asfinag.at/toll/toll-for-hgv-and-bus.

81